AF725851

SAE EDGE™
RESEARCH REPORT

Unsettled Topics Concerning Automated Driving Systems and the Transportation Ecosystem

Dr. Rahul Razdan
Advanced Mobility Institute, Florida Polytechnic University

EDGE DEVELOPMENT TEAM

Jim Boxold, *Florida Secretary of Transportation*

J.W. Taylor, Esq., *Transportation and Logistics Attorney*

Richard Watts, Sr. *VP Progressive Insurance* (retired)

Dale Neef, *Expert in Local Government Planning for "Smart Cities"*

Gary Ralston, *Managing Director of SVN (commercial real-estate)*

Pankaj Mayor, *Marketing Expert, ex-CMO of public software companies*

SAE INTERNATIONAL®
Warrendale, Pennsylvania, USA

About the Publisher

SAE International® is a global association of more than 128,000 engineers and related technical experts in the aerospace, automotive, and commercial-vehicle industries. Our core competencies are lifelong learning and voluntary consensus standards development. Visit sae.org

SAE EDGE™ Research Report Disclaimer

SAE EDGE™ Research Reports focus on topics that are dynamic, in which knowledge is incomplete, and which have yet to be standardized. They represent the collective wisdom of a group of experts and serve as a practical guide to the reader in understanding unsettled subject matter. They are not meant to provide a recommended practice or protocol. The experts engaged have contributed their own thoughts and points of view, and these are not the positions of the institutions or businesses with which they are affiliated. A professional writer has collectivized their input; there is no one contributor's perspective being advanced but rather that of a community of practitioners. SAE EDGE™ Research Reports are the property of SAE International and SAE alone is responsible for their content.

About This Publication

SAE EDGE™ Research Reports provide state-of-the-art and state-of-the-industry examinations of the most significant topics in mobility engineering. SAE EDGE™ contributors are experts from research, academia, and industry who have come together to explore and define the most critical advancements, challenges, and future direction in areas such as vehicle automation, unmanned aircraft, IoT and connectivity, cybersecurity, advanced propulsion, and advanced manufacturing.

Related Resources

SAE MOBILUS® Automated & Connected Knowledge Hub
http://sae.org/Mobilus/Automated

SAE Team

Frank Menchaca, Chief Product Officer
Michael Thompson, Director, Standards, Information and Research Publications
Monica Nogueira, Acquisitions Director
Jill Leonard, Product Manager
William Kucinski, Managing Technical Editor

Copyright © 2020 SAE International. All rights reserved.

No part of this publication may be reproduced, stored in a retrieval system, distributed, or transmitted, in any form or by any means without the prior written permission of SAE International. For permission and licensing requests, contact SAE Permissions, 400 Commonwealth Drive, Warrendale, PA 15096-0001 USA; e-mail: copyright@sae.org; phone: 724-772-4028; fax: 724-772-9765.

Printed in USA

Information contained in this work has been obtained by SAE International from sources believed to be reliable. However, neither SAE International nor its authors guarantee the accuracy or completeness of any information published herein and neither SAE International nor its authors shall be responsible for any errors, omissions, or damages arising out of use of this information. This work is published with the understanding that SAE International and its authors are supplying information, but are not attempting to render engineering or other professional services. If such services are required, the assistance of an appropriate professional should be sought.

ISSN 2640-3536
e-ISSN 2640-3544
ISBN 978-1-4686-0112-1

To purchase bulk quantities, please contact: SAE Customer Service

E-mail: CustomerService@sae.org
Phone: 877-606-7323 (*inside USA and Canada*)
724-776-4970 (*outside USA*)
Fax: 724-776-0790

Visit the SAE International Bookstore at books.sae.org

About the Editor

Dr. Rahul Razdan is Senior Director of Special Projects at Florida Polytechnic University in Lakeland, Florida. In this role, he serves as advisor to the president of the university and helps to manage the Advanced Mobility Institute, an applied research center for the development and testing of autonomous vehicle (AV)-related technology. Named on more than 24 issued patents, Razdan earned a Doctorate (PhD) in Computer Science from Harvard University and both a Master of Science in Computer Engineering and a Bachelor of Science in Electrical and Computer Engineering from Carnegie Mellon University.

Razdan boasts 35 years of experience in startups, academia, and Fortune 500 companies working in such areas as science, technology, engineering, and mathematics education; AV technology; and semiconductor design. The Association for Computing Machinery and the Institute of Electrical and Electronics Engineers have recognized him with “Hall of Fame” inductions. Razdan has also won numerous best-paper awards and has led several successful startups in areas such as electronic design automation, wireless power, artificial intelligence (AI), and machine learning.

contents

© 2020 SAE International. All rights reserved. No part of this publication may be reproduced, stored in a retrieval system, or transmitted, in any form or by any means, electronic, mechanical, photocopying, recording, or otherwise, without the prior written permission of SAE.

Unsettled Topics Concerning Automated Driving Systems and the Transportation Ecosystem

Abstract

Over the last 100 years, the automobile has become integrated in a fundamental way into the broader economy. A broad and deep ecosystem has emerged, and critical components of this ecosystem include insurance, after-market services, automobile retail sales, automobile lending, energy suppliers (e.g., gas stations), medical services, advertising, lawyers, banking, public planners, and law enforcement. These components - which together represent almost $2 trillion of the U.S. economy - are in equilibrium based on the current capabilities of automotive technology. However, the advent of autonomous vehicles (AVs) and technologies like electrification have the potential to significantly disrupt the automotive ecosystem. The critical cog governing the rate and pace of this shift is the management of the test and verification of AVs.

In this SAE EDGE™ report, six senior industry leaders in the impacted ecosystems essay articles which describe sectors of the current automotive ecosystem and the manner in which AV technology can potentially reshape them - providing a mosaic of the massive infrastructure shifts which will be required to absorb AV technologies.

NOTE: SAE EDGE™ Research Reports are intended to identify and illuminate key issues in emerging, but still unsettled, technologies of interest to the mobility industry. The goal of SAE EDGE™ Research Reports is to stimulate discussion and work in the hope of promoting and speeding resolution of identified issues. SAE EDGE™ Research Reports are not intended to resolve the issues they identify or close any topic to further scrutiny.

DR. RAHUL RAZDAN

Advanced Mobility Institute, Florida Polytechnic University

Edge Development Team

Jim Boxold,
Florida Secretary of Transportation

J.W. Taylor, Esq., *Transportation and Logistics Legal Specialist*

Richard Watts, Sr.
VP Progressive Insurance (retired)

Dale Neef, *Author and Expert in Local Government Planning for "Smart Cities"*

Gary Ralston, *Managing Director of SVN (commercial real-estate)*

Pankaj Mayor, *Marketing Expert, ex-CMO of public software companies*

ISSN 2640-3536

Introduction

State of the Industry

Since the 1950s, computing solutions have fundamentally shifted major parts of the world economy. As shown in Figure 1, the first wave of electronics consisted of centralized computing, and the leaders in the field included companies such as IBM, Digital Equipment Corporation, Wang, and others. Fundamentally, these technologies provided productivity solutions for the General and Administrative functions for global business. With this shift, the finance, human resource, and administrative functions of global business were disrupted: Gone were the days of a sea of administrative assistants doing paperwork.

The next wave consisted of edge computing devices such as personal computers, cell phones, and tablets. Companies such as Apple, Amazon, Facebook, and Google added enormous productivity to the advertising and distribution function for global business through these devices. Suddenly, one could instantly reach any customer around the world. This mega-trend has fundamentally disrupted markets ranging from media to retail.

In both cases, the world existed in a previous equilibrium and these economy-shaping technologies disrupted the equilibrium in multiple dimensions (Figure 1). Along the way, employment, taxation, public planning, and legal frameworks had to adjust to the new normal.

We are now at the beginning of the next major disruptive cycle caused by computing. This cycle consists of embedded sensory devices or Internet of Things (IoT), local Artificial Intelligence (AI) systems or machine learning, and global intelligence or cloud resources. Broadly called AI/IoT, these three technologies will disrupt nearly every market segment which uses intelligent mobility solutions. Our current transportation system is the largest to be disrupted and consists of

1. **Automotive companies** with original equipment manufacturers (OEMs) such as Ford Motor Company and Volkswagen; distribution companies like AutoMax Auto Group; and maintenance companies like Jiffy Lube International
2. **Rental market** companies ranging from Avis Rent A Car to local cab companies
3. **Automotive financers** and banks like Bank of America and others
4. **Automotive insurers** such as Progressive Corporation, GEICO, and others
5. **Energy providers** throughout the entire fuel supply chain (e.g., Exxon Mobil Corporation)
6. **Media companies** in radio and billboard advertising like Lamar Advertising
7. **Public sector** including infrastructure (federal, state, and local), taxation, and traffic enforcement
8. **Accident support** organizations that provide legal resources, aftermarket repair, and healthcare services

FIGURE 1. **Economy-changing technologies.**

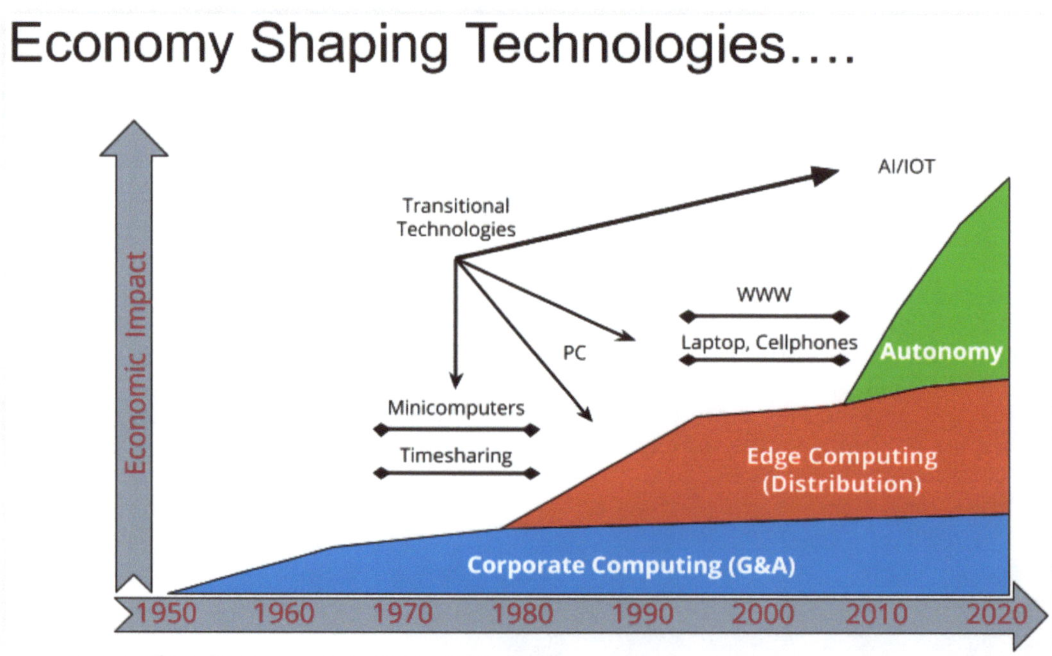

© 2020 SAE International. All Rights Reserved.

TABLE 1. Automotive ecosystem [2].

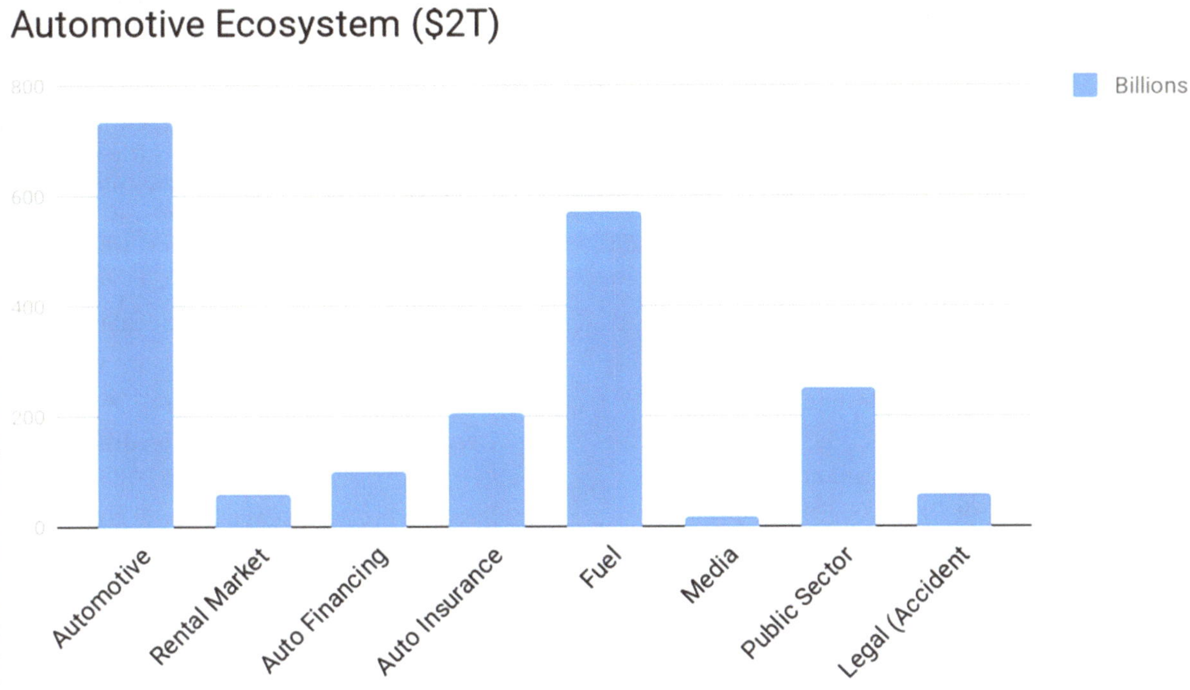

© 2020 SAE International. All Rights Reserved.

Currently, this massive ecosystem operates with four fundamental assumptions:

1. **Ownership**: The vehicle ownership model is an individual.
2. **Human Driver**: A driver is a human with associated liability and insurance.
3. **Fuel Source**: Gasoline is the fundamental fuel source.
4. **Infrastructure**: The public sector provides infrastructure (today: roads and traffic management).

With the advent of AI/IoT, every one of these fundamental assumptions is being challenged and a $2-trillion market is about to be disrupted (Table 1). However, the shape of this disruption is not yet clear and will be highly dependent on the demonstration of robust capabilities of AV solutions, raising the questions regarding AV technology and its impact on current sectors within the automotive industry. The test and validation framework will be central to the type and level of change. Our first report on SAE EDGE™ Test and Validation [1] highlighted these challenges. In this report, we will explore the potential impact of this technology shift with leaders in each of these fields and seek two fundamental questions:

1. What is the current equilibrium in the automotive ecosystem?
2. What are the open questions posed by AV technology from their perspective?

Helping us navigate these topics are

- **Public Policy (Jim Boxold):** Jim Boxold was recently the Secretary for the Florida Department of Transportation where he was responsible for managing a $10-billion budget and over six thousand employees. Jim explores the current automotive ecosystem and the impending benefits as well as challenges of AV technology from the perspective of a state entity.
- **Legal Frameworks (J.W. Taylor, Esq.):** Transportation attorney JW Taylor Esq is a sought-after advisor in the emerging relationship between technology and transportation. J.W. is engaged with the Florida Trucking Association, Blockchain in Transportation Alliance, and mobility-as-a-service (MaaS) providers in their efforts to navigate through the legal framework. He will discuss the current legal frameworks in automotive and open issues for AVs.
- **Insurance (Richard Watts):** Richard Watts was the group president of sales and service for Progressive Corporation and sits on the board of Farmers Insurance Group. From his perspective, he will discuss the current structure of the auto insurance industry and the open issues with AVs.
- **Smart Cities (Dale Neef):** Dale Neef is an author, speaker, and strategic technology consultant who advises organizations and communities on “Smart

Cities," AV infrastructure, municipal broadband, and other digital economy issues. He will discuss the role of local government today in the automotive ecosystem and the potential planning required for the future.

- **Commercial Real Estate (Gary Ralston):** Gary M. Ralston, CCIM, SIOR, SRS, CPM, CRE, CLS, CDP, CRX, CRRP, FRICS, is Managing Director and Partner of SVN Saunders Ralston Dantzler Realty, LLC. Gary is recognized as the most accredited commercial real estate practicing professional in the nation. Gary will discuss the current connection of the automotive ecosystem to commercial real estate and explore the impact of AV technologies in that marketplace.
- **Marketing and Advertising (Pankaj Mayor):** Pankaj Mayor has held executive roles leading marketing at several Fortune 500 firms such as Ansys and Cadence Design Systems. He will discuss the impact of AVs on the current models in media for marketing and advertising.

Autonomous Vehicles and Public Policy by Jim Boxold

Current Automotive Ecosystem

Today, federal, state, and local entities cooperate to build a functioning transportation system which optimizes safety, traffic flow, economic development, and revenue to fund projects. Traditionally, the federal government has managed the automobile, the state has regulated the driver, and with the local government all three collectively support the transportation infrastructure, including

1. Traffic Management: All three government levels use road design, signage, and traffic lights to manage the flow of traffic.
2. Traffic Optimization and Mitigation: Overlaying the civil infrastructure is an electronics infrastructure consisting of various sensors (camera, fog, etc.) with central operations in order to optimize traffic and react to accidents.
3. Accident Support: Law enforcement, emergency, and transportation staff cooperate to manage accidents while federal agencies engage in discovering and resolving systemic issues.
4. Economic Development: Economic development is a major driver for investments in civil infrastructure.
5. Revenue: Large investments in transportation infrastructure are funded with a range of sources including general tax appropriations (federal, state, local), toll fees, and fuel taxes.

States regulate driver licensing, traffic law enforcement, and driver safety records. When a vehicle accident occurs, the first engagement point is local law enforcement which performs an investigation. In the process, they generate a somewhat standardized accident report which includes the following information:

- Date, time, and location of accident (weather conditions may be included)
- Driver(s) and vehicle identifying information
- Description of accident based on driver(s), passenger(s), and other witness reports
- Location of damages and medical issues

After an accident, a well-honed system of insurance companies, lawyers, and the court system arbitrate liability issues between the parties. If the driver is at-fault, the state may revoke driving privileges from the driver based on safety concerns. When the fault is with the vehicle, a similar process involving the National Highway Traffic Safety Administration (NHTSA) occurs with the vehicle manufacturer [3]. NHTSA is responsible for the licensing of vehicles and uses databases such as the Fatality Analysis Reporting System to aid in this task. In its activities, NHTSA is seeking to find and fix systemic vehicle issues and provide feedback on safety issues for vehicles. NHTSA investigations are often triggered by consumers through the Office of Defects Investigation or Crash Investigation Sampling System.

Since 2012 about 92% of new light automobiles are voluntarily equipped with an Event Data Recorder (EDR). This technology is analogous to the flight data recorder in planes; however, the Federal Government has yet to require that all vehicles must have this device installed. In 2010, the NHTSA produced an official regulation which defined the EDR and all the parameters which must be recorded by it when it is available. Examples of these parameters include longitudinal and latitudinal speeds and accelerations, engine throttle and revolutions per minute, time, anti-lock braking system activity, airbags deployment status, roll angle of the vehicle, occupant position, etc. Most of these parameters are recorded until a fraction of a second before an event or accident. Finally, after the accident, a well-structured interview process is used to provide more insight.

Autonomous Vehicle Challenges

With most accidents caused by human error, the promise of AVs is to significantly reduce traffic fatalities. For traffic congestion, the increased level of control of the automobile allows transportation engineers more tools to reduce traffic congestion. As the next mega-market mover, AV development, production, deployment, and service offer the next big growth market, and the greater efficiencies enabled by AV technologies should add value to the whole economy. However, given the size and scope of the current public sector ecosystem, significant challenges exist. The major areas of focus include

- **Regulations:** The assumption that a human driver is responsible for vehicle operation is embedded in statutes and regulations. The fundamental legal framework for testing and deploying AVs cannot exist if these statues are not updated. Some states, such as Florida, have taken proactive action with bills that remove assumptions of a human driver.
- **Safety and liability:** A delicate balance between encouraging risk-taking in development and maintaining safety must be managed. A number of states allow AV testing when matched with high insurance thresholds. As an example, Florida has specified $1-million policy before testing is allowed.
- **Accident processing:** An updated accident form and associated training which replaces the driver interview with more direct prospective observations from law enforcement is required. Since a single AV accident can represent a defect in a large number of vehicles, the path to finding and resolving systemic issues must be much more aggressive. The California Department of Motor Vehicles began adopting regulations for testing AVs in 2014 with an important update in 2018 which requires manufacturers to submit a report within 10 days of an accident. Additionally, California AV testing regulations require that every manufacturer, if authorized to test their AVs on public roads, submit a yearly report that lists all the disengagements of the AV technology during testing.
- **Infrastructure investment:** The fundamentals of AV integration are still fluid; however, deep investments may be required for providing situational information through Vehicle-to-Infrastructure (V2I) communication systems. For infrastructure providers, this fluidity creates significant challenges regarding the timing of investments relative to market readiness.
- **Revenue:** AV technology may reduce the total number of automobiles required on the roads, which may impact vehicle tax revenues. Additionally, revenue connected with fuel sources such as gasoline will be impacted by the shift toward electrification. With the advent of the electronic technology revolution, many powerful tools (e.g., micro-tolling) and traffic management models will become available to the transportation ecosystem. These include

    1. **Adaptive and responsive transportation systems** could enable transportation systems which manage shared resources and monitor, evaluate, and self-manage in a manner similar to computer networks.
    2. **Market-based optimization of the transportation network** could shift the current transportation infrastructure's "peak capacity utilization" design factor to a "demand-supply" structure that uses the demand in the market with quality of service parameters (e.g., trip time) for economic optimization.

Recommendations

Public policy makers must diligently manage safety concerns while building sufficient legislative infrastructure to allow for the development and early deployment of this ground-breaking technology. Critical needs include

1. Scenario Language and Management: How can one clearly define the scope of operation? How do real-world accidents migrate into the test framework?
2. Systemic Risk: What are the technical means by which escalation for system risk can be determined?
3. Automotive Ecosystem: How does one determine the completeness, adaptiveness, and robustness of the infrastructure?

Overall, the science for test and validation is critical because it forms the basis around which regulatory constructs can be built.

AVs and Legal Frameworks by J.W. Taylor, Esq.

Introduction to Historical Automotive Legal Exposure

As stated earlier, "over the last 100 years, the automobile has become integrated in a fundamental way into the broader economy." The fundamental integration must include integration of the automobile into the legal framework that supports the automotive technology. One may assume that the same will be true of the legal framework in the emerging AV technology. In fact, soon after the invention of the "car," the legal environment was invited to the first crash(es). There are varying stories, but the first "accident" may have occurred

FIGURE 2. **Accidents, insurance, and the law.**

Unitone Vector/Shutterstock.com

in 1869, when Irish scientist Mary Ward was thrown from her steam-powered automobile while taking a curve too fast. Ms. Ward was thrown under the automobile and was killed instantly [4].

The first automobile accident involving a pedestrian came in 1896, when Bridget Driscoll stepped from a curb in London into the path of a gas-powered automobile driven by Arthur Edsall. The top speed of the automobile was only 4 mph. While Mr. Edsall was arrested, the jury returned a verdict of "accidental death." The coroner was quoted as saying he hoped "such a thing would never happen again."

Of course, and unfortunately, Bridget Driscoll was not the last pedestrian involved in an accident with the automobile, nor is she the last to be involved with an accident related to emerging automobile technology. Fast-forward to the night of March 18, 2018, in Phoenix, Arizona, Elaine Herzberg pushed her bike across the dimly street, 360 feet south of the crosswalk, and abruptly stepped from the median and into the path of an "autonomous" Uber vehicle. Interestingly, at the inquest Mr. Edsall received a verdict of "accidental death" following six hours of jury deliberation while Ms. Herzberg's case was resolved in settlement before litigation commenced [5].

Present Legal Framework

From then until now, one needs to only glance outside the traditional automobile to see billboard after billboard advertising lawyers for automobile accidents. Television advertisements are just as plentiful. Current automotive accident case law continues to evolve differently in each state; however, it remains rooted in driver negligence.

The essence of a "driver negligence" claim is that the injured party must prove (1) the at-fault party had a duty to use reasonable care to prevent injury to others, (2) the at-fault party breached that duty, (3) such a breach was the actual and proximate cause of the other party's injuries, and (4) the other party suffered damages [6]. Although each driver negligence case is unique to the specific facts of a case, all cases share one factor: The plaintiff must prove that the defendant (commonly the driver of the vehicle) acted negligently and that this negligence caused specific damage. This is easier to accomplish in certain cases, such as when an intoxicated driver causes an accident. In other situations, though, proving negligence may require substantial evidence and expert testimony.

AVs add a significant layer of complexity to the negligence framework. When the controls move from a traditional driver to a third party - whether a manufacturer, a software provider, or an off-site controller - the liability shifts as well. Beyond a responsible driver at the wheel, other parties owe a duty of care in placing a vehicle on the road that will operate safely, and oftentimes will be found partially responsible for damages.

Each party's exposure is different across the states. Some states even find the level of responsibility can be lowered due to the plaintiff's own negligence. For example, Florida has adopted a pure comparative fault theory, which means if a plaintiff is at fault, that percentage will diminish proportionately the amount he is entitled to recover, while Maryland has adopted a pure contributory negligence theory which bars recovery if the plaintiff is found to have contributed to his damages [7]. These theories of law can show a framework of how other parties outside the "negligent driver" may be found to have been partially responsible for damages and injuries sustained in an accident involving AVs.

Emerging Liability Structures

Even with well-established personal and commercial automotive law, AV technology raises more questions regarding liability than answers. Who is responsible?

- **Car manufacturers, including software providers**: Today, nonautomated vehicle manufacturers can be found liable for damages under product liability. Will manufacturers of AVs be liable for defects? Will they be found liable under warranty?
- **Driver who was not "driving"**: Traditional notions of driver negligence may be difficult to apply to the person who is not involved in operating the AV. Can a driver be negligent if they are not actually operating a vehicle? What is their duty owed to other vehicles on the road?
- **Local governments**: AVs need smooth, well-maintained roads with clear signage and visible road markings. What happens if the local governments fail to implement the necessary changes? Will they be shielded from liability by sovereign immunity?
- **Monitors or operators outside the AV**: Will remote operators be found negligent much like current drivers of nonautomated vehicles?

The very essence of a vehicle negligence claim involves a "reasonable person" analysis, typically a "reasonable driver." However, in a self-driving vehicle, there may not be a "driver" by which to measure against the reasonable person standard. So, what is the theory of law in the case of a vehicle accident involving self-driving vehicles?

Negligence and Negligent Entrustment of Vehicles

Negligence claims could be directed toward operators of AVs, whether they be the driver in the vehicle or the remote operator in a separate location; however, it may be difficult to prove the driver or remote operator acted negligently if control over the AV is limited. The point of automating a vehicle is to try to eliminate the human error that is often the basis for automotive negligence claims.

Similar issues are present in negligent entrustment situations, where the owner of a vehicle allows another party to drive or operate the vehicle who is then involved in an accident with that vehicle. In a negligent entrustment case, the injured person must prove the following:

1. The owner entrusted the vehicle to the driver.
2. The driver was negligent.
3. The owner knew or had reason to know that the driver had no driver's license, was incompetent, or was known to drive recklessly.
4. The driver was negligent at the time the incident occurred.
5. The driver's negligence caused the incident.

Again, this theory of law can be applied to both the "driver" and/or the operator of an AV, but may be more difficult to prove negligence of "drivers" and operators when they may not be in control of the AV.

Res Ipsa Loquitur and Products Liability

When faced with the challenges of determining negligence, courts may look to use the theory of "res ipsa loquitur." The basic elements of res ipsa loquitur are (1) the event must be of a kind which ordinarily does not occur in the absence of someone's negligence, (2) it must be caused by an agency or instrumentality within the exclusive control of the defendant, and (3) it must not have been due to any voluntary action or contribution on the part of the plaintiff.

This theory removes the necessity of proving negligence and applies stricter liability. If the courts follow this path, it may cause many AV owners to seek out opportunities to collect from AV manufacturers due to products liability, requiring them to prove (1) the product was defectively designed so as to render it unreasonably dangerous, (2) a safer alternative design existed, and (3) the defect was a producing cause of the injury for which the plaintiff seeks recovery.

With many of today's nonautomated vehicles already having so much software and automated parts, it seems this theory may be applicable to AVs in much the same way as it is already applied to vehicles openly today.

Breach of the Implied Warranty of Merchantability

Another legal option that could be used against manufacturers and software programmers is the breach of the implied warranty of merchantability. To establish a claim for breach of the implied warranty of merchantability, the injured party must show that the merchant sold a good which did not meet one of the standards of merchantability, an approach applied to cases involving nonautomated vehicles:

1. Passes without objection
2. Is of fair average quality
3. Is fit for ordinary purposes
4. Is of even kind, quality, and quantity
5. Is adequately contained, packaged, and labeled
6. Conforms to promises on the packaging

For the same reasons as above, this theory may apply to cases involving AVs based on existing cases dealing with nonautomated vehicles today.

Liability of Local Governments AVs need appropriate infrastructure to operate effectively, and local governments may need to ensure that infrastructure is updated and well maintained. Currently, local governments are rarely found liable in the case of vehicular collisions, but what happens when AVs cause a collision due to a faded street line or an unpatched pothole?

Government entities are often protected by the theory of sovereign immunity in cases involving collisions and defects in public infrastructure. However, there are three ways to avoid a sovereign immunity defense [8]: (1) if the entity fails to correct or warn against a known dangerous condition, (2) the entity fails to maintain existing roads, or (3) the entity fails to properly construct or install roads.

Even though these issues have not been litigated specifically with AVs in mind, local government should start discussing what they need to do to shield themselves from liability. AVs are coming and they are predicted to make up 95% of all new vehicles sold by 2040. Considering the case law limiting sovereign immunity, local government should take steps to be prepared.

With all these theories of law that could be applied, which of these theories are likely to be the prominent theory when dealing with accidents involving AVs? It will be no surprise if the answer is "all the above." For now, there is little to glean from the actual incidents involving AVs, whether they be nonfatal or fatal cases. There is little precedent on AVs specifically, but there may be some guidance gleaned from existing cases and law involving nonautomated vehicles, manufacturers, and other operators.

For example, in California, manufacturers filed 30 AV accident reports between 2014 and mid-2017. Only two were caused by the AV. Interestingly, most of these crashes were rear-end crashes caused by the traditional vehicle [3]. Presently, the laws in most states have a presumption of negligence against the automobile who rear-ends the other. Is the present understanding of rear-end crashes the reason? That being said, none of these incidents have set any precedential value.

To date, there are five known fatalities (one Level 3 fatality [pedestrian], four Level 2 fatalities [all drivers]) involving AVs. Much attention was given in the media to the 2016 fatality in Florida involving a Tesla AV. It appeared the Tesla AV, while in autopilot, failed to sense a tractor-trailer crossing in front of it due to the "white sky" background. The Tesla driver was killed. Again interestingly, the NHTSA found the driver remained responsible for the "ultimate control." There was no civil litigation in the Tesla AV case. Will "ultimate control" of the AV be the basis for determining liability? There seems to be some traction around this idea.

In March of 2019, a second fatality involving a Tesla AV occurred in Delray Beach, Florida. In this incident, a Tesla Model 3 struck a tractor-semitrailer combination vehicle that crossed in front of the AV to get to the opposite side of a divided highway. The Tesla drove underneath the semitrailer, and the driver died as a result. Data from the Tesla's computer shows that the driver engaged the vehicle's Autopilot roughly 10 seconds prior to the collision. All aspects of the crash, including the actions of the Autopilot, the driver, and the driver of the tractor-trailer remain under investigation (NTSB Preliminary Report #HWY19FH008).

The Phoenix fatality was alluded to previously. The incident occurred in Arizona in 2018. One year before the incident occurred, in March of 2019, the Yavapai County Attorney's Office found no criminal liability on the part of Uber but allowed investigators to continue looking into whether the human operator inside the AV at the time of the crash was negligent. The National Transportation Safety Board and the NHTSA turned the investigation over to the local police.

As of July 2019, the Maricopa County police department was still investigating the crash in order to decide whether or not to bring manslaughter charges against Rafaela Vasquez, the safety driver who was supposed to take manual control of the Uber in the event of an emergency. At the time that Herzberg was killed, Uber had disabled the emergency braking capabilities on the vehicle to smooth out the ride, and Vasquez was, allegedly, not paying attention as she was streaming a television program on Hulu on her cell phone.

Another fatality occurred in March of 2018 involving a Tesla AV that drove into a highway safety barrier, killing the driver. This incident is ongoing and may lead to litigation. The attorneys for the family have alluded to including legal theories based on "product liability, defective design, failure to warn, breach of warranty, intentional and negligent misrepresentation, and false advertising" [9]. Seemingly, but maybe not likely, the only legal theory missing is "negligent entrustment." For now, however, the current legal framework can only speculate on the emerging tort theories.

Emerging Tort Defenses Researchers estimate that AVs can reduce accidents by up to 90 percent. That equates to saving more than 30,000 lives each year, not to mention injuries to body and property [10]. This is because you remove the "human error" factor from driving that are found to cause so many accidents, injuries, and deaths per year. As GM Chairman Bob Lutz put it, "The autonomous car doesn't [...] do drugs, doesn't text [...] and they don't go to sleep."

Notwithstanding, can a manufacturer be subject to liability even where the AV technology did not "cause" the "harm?" As we discussed earlier, the theory of negligence requires a party to have acted "negligently." In order to act "negligently," you must have control over such actions. So who has the control in an AV? The answer to that is the typical answer every lawyer is known for, "it depends." Each incident will be very fact intensive. For example, will an intoxicated "driver" act negligently by "operating" an AV when the AV is supposed to have the technology to drive itself? Or, in that same scenario, will the manufacturer be found to have acted negligent for not creating software that can operate appropriately even with an intoxicated operator? And if not, has the manufacturer adequately warned that the vehicle must be operated by a nonintoxicated person? Would there be a "negligent entrustment" issue for AV owners to allow "drivers" not properly trained to "drive" AVs?

Since there appear to be more questions than answers with little to no precedent to draw from, may we look at analogies? Some have looked to case law related to elevators or autopilot systems for direction. The most intriguing analogy, however, has to be comparing AV accidents to those caused by "equine transportation." Simply put, both horses and AVs have certain "intelligence" yet can misunderstand the surroundings and maneuver independently of the driver's will. In certain areas, the owner and/or rider of the horse may be strictly liable for any injury or damage caused by a horse coming into unexpected contact with a motor vehicle due to specific laws that limit the areas where horses may be present. For example, certain jurisdictions may not permit horses to roam or be ridden freely within a certain distance of a roadway. Still most jurisdictions follow some type of negligence framework whereby the owner, rider, and/or keeper of the animal may be held liable. In these circumstances, the failure to equip their horses with proper riding instruments, maintain proper enclosures, or treat horses that are prone to frightening easily with due care will be found to be negligent and, therefore, held liable. There are several recent examples of motor vehicle accidents, however, where owners/riders were not held liable due to the courts finding that the behavior of the horse itself was outside of the reasonable control of the human "in charge" of the horse. In a 1997 case out of New York, a racehorse, that was accustomed to the noise and sight of vehicular traffic, was being unloaded from a trailer about 20 feet from a highway when it suddenly became frightened and ran into the road. Nobody could explain why the horse spooked, and its handlers were not held liable for its behavior [11].

Can a horse or an AV truly "understand" its environment? What then do we do with liability when "horse power" acts "independently?" Can this be defended as not driver negligence?

Presently, car manufacturers do not have a duty to design an accident-proof car or to make their vehicles safer where the danger to be avoided is obvious. The intended purpose of an automobile is to drive across terrain, and not to be involved in collisions with other objects, despite the manufacturer's ability to foresee the possibility that such collisions may occur.

Even though a vehicle's purpose is not to be involved in a collision, the courts have stated it is foreseeable and therefore a manufacturer has a duty to maintain a level of safety of each vehicle. A manufacturer owes a duty to use reasonable care in the design of its products to protect against an unreasonable risk of injury or enhancement of injury to a user of the product. "[I]t is the obligation of an automobile manufacturer to provide more than merely a movable platform capable of transporting passengers from one point to another. The passengers must be provided a reasonably safe container within which to make the journey."

The question of manufacturers and their liability in AVs may be tested under this duty. However, the duty for nonautomated vehicles is based on the idea that collisions are foreseeable and yet the idea behind AVs is that collisions would not be as prevalent. If that is the case, will the manufacturers still be found to be at fault like they have in the past for incidents involving collisions?

Manufacturers may also have a "state of the art" defense. Some jurisdictions have laws that provide an affirmative defense to a product's liability action if the product is "state of the art" [12]. The purpose of a "state of the art" defense is to protect a manufacturer from liability for failure to anticipate safety features that were unknown or unavailable at the time a product was manufactured and distributed. The defense may be asserted against a manufacturing defect claim, and may be a complete defense in product defect cases. This defense could help protect manufacturers as AVs begin to break into the market. However, it will not protect them forever and manufacturers should be cognizant of the limitations of their defenses.

Recommendations

The nature of the American legal system is that the legal theories for liability will be tested in the courts with a series of cases based on the current legal frameworks with a bias toward the "reasonable human" guidance. The following actions would greatly aid this dispute resolution process to build a new area of law which can function with stability and creditability.

1. Regulatory Backdrop: To move away from the "reasonable human" guidance, there is a need for a regulatory backdrop which can set the reasonable expectations for AVs.
2. Safety Science Backdrop: As mentioned, the regulatory backdrop needs improvements in the science for safety, and by extension, any future legal theory of liability will require the same.

Overall, today's society tolerates other humans on the roads even with the safety results to date because there is a reasonable expectation of the behavior (even bad) of humans. To tolerate any level of risk, AVs will have to build a reasonable expectation of behavior which is easily communicated to their surroundings.

Considerations for the Automobile Insurance and Autonomous Vehicles by Richard Watts

Current Automotive Ecosystem

The current state of the $200-billion automobile insurance industry consists predominantly of sales to the consumer, either directly or through intermediaries and the subsequent risk assessment of human drivers and their vehicles. As one of the original consumers of massive amounts of data, insurance companies have used that information in the underwriting process. Much like modern machine learning or AI, insurance companies collect accident data and find correlations to manage risk. Historically, the most important factor has been driver history which leads to the "points" system.

The next leap in insurance underwriting was the usage-based model where drivers could share driving behavior to reduce insurance costs. In this model, driver behavior can be tracked in a fine-grained manner and this data can lead to better underwriting solutions. In terms of technology, Table 2 shows some of the safety technology pieces whose impact is folded into the underwriting process. The technologies can be separated into three categories - driver safety, driver fail-safe, and driver augmentation. Driver safety technologies aid the human driver, so the underwriting treatment is just reflected in the accident data. Driver fail-safe systems are tools used to detect and manage specific causes of unsafe behavior. The most interesting of these are the driver augmentation technologies which start entering the gray area in terms of driver underwriting and automobile algorithm risk assessment.

AV Challenges

The Advanced Driver Assistance Systems (ADAS) are forming a shift toward AVs which start disrupting the automobile insurance industry. In the longer term, the major factors in the disruption include the following factors:

1. Size of Business: If AVs can reduce accidents by 90 percent, what does it mean for the demand for insurance?
2. Fleet versus Individual: If AVs drive a shift from individual ownership to mobility-as-a-service, will the insurance business shift from consumer to commercial fleets. Will fleet operators just self-insure?
3. Original Equipment Manufacturer (OEM) Insurance: If the driver is an AV, will OEMs need

TABLE 2. Safety technologies.

Driver safety technologies	Driver fail-safe	Driver augmentation
Night vision	Alcohol ignition	Auto lane centering
Blind-spot detection	Driver drowsiness detection	Adaptive cruise control
Vibrating seat		Collision avoidance
Sat/Nav traffic warning		
Traction control		
Electronic stability control		
Automated lighting		
Rear cameras		

© 2020 SAE International. All Rights Reserved.

TABLE 3. Summary of structural impact.

Insurance industry structure changes	Insurance capability shifts
Insurer consolidation, staff retraining, massive data storage requirements	Develop cybersecurity capability - cyber theft, ransomware, hacking, and misuse of data
OEMs to become significant partners or competitor to insurers	Develop new product liability products - for software bugs, memory issues, algorithm defects
Continued aggregation of body shops, tow and salvage operations	Develop infrastructure insurance products - cloud server systems, public infrastructure
End of taxi fleet insurers Commercial fleet insurers heavily and impacted first	Develop reinsurance programs for those OEMs and municipalities that elect to self-insure

© 2020 SAE International. All Rights Reserved.

fleet-wide insurance. In addition, given the same algorithm is used in a vast number of cars, how is systemic risk managed?

4. Infrastructure Insurance: If infrastructure will be an important component of the driving task, is there a need for insurance in case of failure?
5. Software Updates: For both OEMs and infrastructure, will insurance sign off have to be connected to the update?
6. Human Driving: Will human driving skills diminish with increased use of AV systems?

The range of risks may well move toward low probability, high-cost events like hurricane insurance. If this is the case, will the industry move toward various levels of reinsurance with a government backstop?

At a more practical level, the process for assessing risk for humans has been honed over nearly a hundred years. What is the practical method by which to assess the risk for AV engines? More information will have to be gathered, and this will mean a significantly updated accident report. The updated AV accident report will link much more deeply into the perception and decision-making systems of the automobile. In addition to "black box" capabilities, it is likely that accident reports will have to include human interviews of the behavior of the AV. Currently, the largest cause of accidents is humans hitting AVs. Finally, given the potential for systemic failure, it is very likely that "near miss" and other related data will have to be captured. All of this means that insurance companies will have to raise significantly their technical competence at all levels of the organization. Currently, ADAS are leaking into the marketplace, and there is limited data upon which to make reasonable underwriting decisions. At a minimum, it is important to capture the engagement of automated systems to start the process of building historical profiles of AV-like systems.

Finally, the insurance framework sits within the construct of a legal framework for AVs which has yet to be developed (Table 3). Through the legislation, regulatory, and court resolution process, important liability constructs will be developed. In these processes, insurance companies will be an important player in determining liability, accident processing, and data availability issues.

Recommendations

Fundamentally, insurance is about data to determine liability and then pricing for risk. Due to the lack of data and immaturity of the technology, state-regulated insurance markets have yet to develop a plan. Several experiments are in flight with concepts such as umbrella insurance for AV OEMs (e.g., Volvo self-insurance).

To make progress,

1. Data Acquisition: Insurance companies need to gain access to data surrounding AV accidents. What data is required? This is gated by the progress on test and validation science.
2. Insurance Market Structure: Given the potential system issues, research is required to examine whether "rare" event markets with high costs (e.g., Hurricane Insurance) might be a better structure for AV Insurance. Does a government backstop make sense?
3. Insurance Capability Shift: Today, the focus of insurance are physical accidents, but in the future, it may well make sense to insure for other events such as cyber attacks or AV "disability."

Smart Cities and Autonomous Vehicles by Dale Neef

It's true that safety standards and use guidelines for AVs will be set by federal and state officials, but just as "all politics are local," connected and autonomous vehicle (CAV) deployment will ultimately depend most on decisions made by town, city, and county government officials. That means that the public sector - particularly at the local level - will be much more involved with AV deployment, and much more influential in the evolution of AV technologies, then they probably realize. They have a significant responsibility for right-of-way and zoning (e.g., whether AVs will be allowed in school zones or where "idle" robo-taxi AVs

FIGURE 3. Commercial real estate and city planning.

Graphic farm/Shutterstock.com

can loiter), funding built infrastructure (e.g., dedicated lanes, curb drop-off and pickup), and authorizing and supervising supportive connectivity (e.g., coordination with advanced traffic management systems, 5G deployment, etc.). This also includes setting pricing requirements and equitable routing, scheduling standards for transit customers, and managing increased congestion, excessive expenditures, poorer public transit services, data privacy leaks, or serious AV-related accidents.

City and county governments also usually have a significant amount of control over local transit authorities and therefore will be seen to have a guiding role in setting pricing requirements and equitable routing and scheduling standards for transit customers as those agencies move to adopt AVs in the future.

And although state transportation departments officials may set broad policy on AVs, it will ultimately be city and county authorities and Metropolitan Planning Organization (MPO) that will be held responsible if local AV policies result in increased congestion, excessive expenditures, poorer public transit services, data privacy leaks, or serious AV-related accidents.

This means that planners, city managers, city engineers, MPOs, and local elected officials need to be thinking now about the implications of CAVs appearing (or not) in their community.

There are a number of policy and infrastructure issues that public sector leaders need to consider.

Built Infrastructure Implications

- **Right-of-ways and dedicated zones (spatial segregation) for AVs and pedestrians**: Although AV sensors, both inside and outside the vehicle, may be effective at preventing collisions and pedestrian-related accidents, separating AVs from other traffic or pedestrians could assure safety and reduce congestion. That can be done in a variety of ways - through dedicated AV lanes, AV-free zones, or AV-only areas - and although state authorities may control major highways, counties and municipalities will often be responsible for traffic within their own areas of jurisdiction. In short, it's often going to be at the county and municipal level that decisions on when, where, and in what form AVs that will be allowed will be made.
- **Street parking and curb access drop-off and pickup**: If AVs take the form of ridesharing, there will be increased congestion along curbs for loading or unloading passengers. Planning departments will need to rethink traffic patterns, redesign access and one-way systems, and reconstruct curbs and street parking areas throughout the municipality.
- **Reallocation and repurposing of parking spaces**: Mentioned previously, if AVs don't need to remain parked all day, parking lots can be repurposed or eliminated entirely, freeing up crucial land for retail or housing space.
- **Electrical recharging framework**: There will be a much greater demand for electric recharging stations - near restaurants, filling stations, grocery stores, and around residential housing. Providing safe and reliable recharging networks - tied to the grid - will often require considerable planning with private utilities and may involve altering local zoning laws.

- **Poles, lights, sensors, and intersections**: Although early AVs will rely predominantly on in-vehicle sensors, as the technology evolves toward vehicle-to-everything (V2X), 5G, and edge computing, external sensors - on poles, signs, curb hardware, or on pedestrians themselves - will become critical. The concern is both functional and aesthetic, and how those sensors are purchased, sited, and maintained will ultimately be the responsibility of the local authority.
- **Connectivity**: That new level of V2X connectivity will be largely dependent upon primary optical fiber availability (gigabyte level), which may mean that for small towns and satellite cities, broadband requirements will grow exponentially in the coming years. As the technology evolves into 5G and beyond, funding, zoning, dig policies, pole right-of-way, and complex procurement contracts (often through public/private partnerships) will all be driven by municipal authorities.
- **Road markings, signals, signs, and maintenance**: It may seem trivial, but simply repainting road markings sooner than expected - or expanding a painted solid road line from 4" to 6" - can result in significant costs to a municipality.

Policy Implications

- **Finance**: Autonomous and electric vehicles will significantly change many aspects of public sector finances - both in terms of revenues and expenditures. On one hand municipalities may see a significant loss of revenues in that traditional sources of municipal funding - traffic citations, parking violations, vehicle property tax, gasoline taxes - will all be considerably reduced with shared, fleet-based AVs and a fewer number of privately-owned cars. Safer AVs should reduce expenditures on emergency services and the police force - on average, 70 percent of a municipality's police activity is traffic related, so AVs could at the same time significantly reduce both incoming revenues and manpower requirements. At the same time, engineering departments will need employees with different, probably more costly, skills (most transportation engineers know about concrete and asphalt - not 5G networks, sensors, or edge computing).
- **AV Infrastructure Costs**: New curbs, signage, dedicated lanes, reconstructed drop-off locations will require extra expenditures. And public transit agency subsidies may need to be rationalized (particularly if costly and inefficient busses are eliminated and replaced with city owned and operated shared AVs). This disruption may push a municipality to rethink policies on congestion charging or other revenue options such as vehicle miles traveled (VMT) fees.
- **Data Collection and Privacy Policies**: Because AVs will potentially generate an enormous amount of data, local authorities will want to create data use policies that ensure a level of privacy for individuals, while at the same time satisfying the need for public collection of "Big Data" for both industry and the public sector. These policies will have to be both principle-based and practical. Potentially sensitive data (e.g., passenger activity) will need to be anonymized and policies put in place - often with the collaboration of private sector companies and technology providers - to safeguard information technology and network security. That collected data is no good if it can't be organized and understood, so analysis tools (transportation demand management, etc.) will need to be made available.
- **Tendering and Procurement**: Some AVs will be privately owned, but many others may be managed by private industry groups (Waymo, Avis, Uber, etc.) in fleets of robo-taxis. Counties and municipalities may want to engage in public/private partnerships with those industry groups and local transit authorities in order to provide the public sector with additional revenue and oversight. Supporting technology and infrastructure projects (5G networks, Advanced Traffic Management Systems) will usually involve tendering with private sector groups. These are key infrastructure projects that will almost always affect other local government agencies - Public Transportation, Emergency Services, Law Enforcement, Public Works, Information Technology - and involve complex tendering and sophisticated program management. This cross-department and strategic tendering is not something that municipalities or counties with multiple department and agency silos have traditionally been very good at, which means they will need to develop more collaborative and flexible procurement programs and contracting methods that ensure interoperability, scalability, and strategic interagency continuity.
- **Codes, Permissions, Zoning**: Possibly most important, municipalities will need to reassess existing laws and municipal codes that affect deployment of AV technologies. A laissez-faire approach to AV use may result in congestion and disruption, while a defensive, restrictive approach may undermine the public and private benefits of an important new technology. Municipalities may restrict AV accessibility in certain neighborhoods, or promote their use in main thoroughfares that lead to business centers, shopping, medical offices, transit hubs, etc.
- **Education and Staffing Considerations**: Whatever the strategy, local governments will require a

dedicated, talented staff to develop, engage, and push through policy changes. That's not really a role that most local authorities (city managers, planners, city engineers, IT directors, economic development directors, etc.) will relish, and few seem to appreciate that these types of infrastructure and policy changes are almost certainly coming with the advent of AV technologies.

It's important to realize that the public today has little appreciation of AV deployment possibilities or their implications - most people who are thinking about it at all, simply assume that the technology will be ADAS+, and mostly privately owned. Fleets of AV robo-taxis, autonomous home delivery, police force reductions, VMT fees, curb infrastructure expenditures are not topics that most people would even associate with self-driving cars. At the same time, as various recent surveys have shown, adverse public reaction to AVs - whether because of safety concerns or the effect of AVs on driver/job elimination, etc. - may be growing.

For its part, the private industry has done very little to promote the evolution to shared vehicles or AVs - and show little sign of being either willing or able (other than through broad concept advertising) to engage with the public on AV issues. Neither federal nor state governments want that role either. That means that it will fall to the local public sector leaders to make the case for action through community-level programs of public outreach - or risk the sort of anti-AV rebellion that has surfaced in early test beds (e.g., Arizona).

Recommendations

For city managers, city engineers, metropolitan planning organizations, and local elected officials, the implications of CAVs appearing (or not) in their community need to get on the radar screen. Fundamentally,

1. Education: CAV technology is quite a step away from the civil and policy roots of most planning functions. An imperative is to build expertise in the organization to understand regulations and potential requirements.
2. Multiuse Functionality: For projects in planning or in flight, some thoughts toward future enablement (e.g., place for fiber) or multiuse capability (e.g., parking garages which can be used for other purposes).
3. Managing Investment and Timing: Before making large investments in CAV technology, it would be wise of local planners to look for stability in federal and state regulations.

Finally, for any government entity, safety is the first concern, so tracking the core test and validation technologies which are likely to underpin regulation as well as requirements for local infrastructure is critical.

Autonomous Vehicles and Commercial Real Estate by Gary Ralston

In most cities, the combination of automotive dealerships, maintenance centers, auto parts stores, parking lots, and fuel facilities is a large driver for commercial real estate representing between $150 and $200 billion of value. NAREIT estimates that retail is about $2.4 trillion of the $16 trillion commercial real estate market in the USA. In addition to the retail automobile market, the commercial marketplace in the form of trucking and warehousing is also a large market connected to the current automotive industry. Given the current footprint, AVs will have a meaningful impact on the commercial real estate marketplace.

Truck Platooning

According to the European Automobile Manufacturers' Association, truck platooning links multiple trucks in a convoy using automated driving support systems. The lead truck - operated by a human driver - effectively sets the pace while trucks in the convoy follow along, maintaining a set, close distance between each other. Currently this program requires follower trucks in the convoy to have human safety drivers onboard, but eventually the follower trucks would be AVs. Platooning optimizes transport by increasing the payload per driver and provides other benefits in terms of fuel savings.

From a commercial real-state point of view, platooning may also allow for more goods to be in transit and thus reducing the time goods spend in the warehouse, potentially reducing the amount of required warehouse square footage. This, in turn, may decrease demand for warehouse space and impact the location and layout of warehouse space (e.g., expansion of parking area to accommodate more containers and trailers).

Car Ownership and Auto Dealerships

Car ownership is primarily a function of allowing people to move from their home to work and other locations. New York City is a good example of the impact of available public transportation on car ownership. According to the NY EDC only 45 percent of all households in the city own a car. And only 22 percent of households in Manhattan own a car. It is interesting to note that although almost half the households in the city own cars, only 27 percent use cars to commute. Transportation needs are supported by about 80,000 for-hire vehicles in New York. The current car ownership rate in the USA is 1.8 cars per household, and if AVs enable a shift in ownership similar to NYC, the impact for commercial real estate would be profound. Will this happen?

"We're moving to a future where people don't own cars," says Dr. Daniel Sperling, director of the Institute of

Transportation Studies at the University of California, Davis. In contrast, Dr. Alain Kornhauser, director of the program in transportation at Princeton University, says "While privately owned cars are unlikely to vanish completely - especially in rural areas, where it might prove inconvenient to wait for a driverless taxi - the number of people who own cars and the number of cars owned per family will drop sharply. While the intensity is unclear, it appears there will be an impact."

The implications for auto manufacturers and auto dealerships are enormous. Automobile dealer (NAICS 4411) sales in 2018 were $1.04 trillion. A significant decline in this industrial sector could be problematic for the retail economy. Also, as to commercial real estate, Statista reports that there are about 37,500 automotive franchise establishments in the USA. Following this argument, there might be eventually a need for significantly less auto dealerships. Most of these properties have good location characteristics and could reasonably be recycled/redeveloped into other commercial/retail uses.

Although AV should result in fewer cars and fewer car dealerships, other industry metrics are probably more correlated to miles traveled. Thus, automotive repair and parts businesses would probably not be impacted as substantially as dealers. The same is somewhat true of gas stations where fuel sales are correlated to miles traveled more than the number of cars.

Parking

With the advent of mobility-on-demand enabled by AVs, the commercial need for parking is likely to be restructured. Private and public entities may experience issues associated with reduced income from parking fees, and associated bonding structures may also be impacted. In addition, parking lots for AVs could be in a less-intensive use geography, i.e., a few blocks away from the destination or even in a fringe area.

Parking garages and parking lots can be redeveloped into more commercially intensive uses. Some real estate developers are already looking ahead, and designing parking garages with the anticipation that the future use will be more consistent with the associated building use, i.e., office, residential, retail, etc. As an example, ceiling heights and other space factors will be designed to conform to the associated use for redevelopment in the future. Finally, AVs may well enable automobile-free zones which can lead to more pedestrian friendly real estate development.

Recommendations

For commercial real estate managers and developers, AV technology is a potential tsunami similar to the impact of e-commerce. E-commerce fundamentally changed the economics of malls, storefronts, and warehouses. Commercial real estate developers would be wise to consider the following:

1. Impact of Current Financial Structures: Many current developments have built bond structures based on current revenue models (e.g., parking). For current and future projects, it would be wise to stress test these models based on the impact of various stages of AV technology.

FIGURE 4. AV marketing and advertising models.

Fmgt/Shutterstock.com

2. Education: Similar to the other players in the automotive ecosystem, education on the potential shift in usage models for AVs is a good first step. For commercial real estate development, the viability of the projects is a function of the penetration rates of AVs into the marketplace which itself is gated by the core science on test and validation of AVs. Thus, tracking the fundamentals of the validation process is a good early indicator of the potential economic impacts.
3. Multiuse Structures: Similar to public sector investors, commercial investors would be wise to explore multiuse concepts to manage financial risk.

Media, Marketing, and Advertising in the Age of AVs by Pankaj Mayor

In the last two decades, there have been two significant shifts in the media and advertising industry. The first was caused by the widespread adoption of the Internet, forcing traditional print media and advertising to evolve into online media. The second was caused by the smartphone, which caused the shift from online media and advertising on the desktop to mobile devices.

The way media and advertising companies serve content now is significantly different than in the twentieth century. It is often instantaneous, personalized to the audience, uses a lot more video, serves up ads that are highly targeted, is very interactive/participative, etc. Companies that did not embrace either shift have struggled, and those that were on the leading edge of the change have been very successful. Today every media company must have a mobile strategy.

The adoption and growth in the use of AVs will create a disruption and an opportunity on an even larger scale than the previous two shifts. The pervasive use of AVs will offer up a captive user, whose location is always known. The environment of the automobile will allow this user to have an immersive experience - whether it is interacting with media or being exposed to highly targeted advertising.

Changes in Advertising

Historically, advertisers have tried to reach people who are in a car with billboards and radio advertising and hope that the driver and passengers will remember the advertisement. Video was not an acceptable medium, as it was perceived as a distraction and a safety hazard. Advertisers had no way to target a user based on their profile or preferences - reducing the effectiveness of the advertisement.

Autonomous cars will be a game changer for advertisers. Radio and billboard ads can be streamed to a screen in the car and the passengers can purchase the item online right away. The car can take a detour and drive the passenger to pick it up. An increase in autonomy and connectedness of the car is creating a much greater awareness of driver location and driver identity. With proper privacy safeguards in place, this creates an opportunity for advertisers to refine their targeting of the consumers of their products. For the customer it is also helpful as it reduces clutter and provides them information that is more relevant.

AVs as the New Media Platform

Americans spend an average of 300 hours in their cars each year with 90 percent of them being over the age of 16 - making it a very attractive demographic to advertisers. As AVs turn drivers into passengers, this presents an opportunity for advertisers to target potential customers over many new viewing hours.

As cars become autonomous, many of the functions that the driver performs will be obsolete, and these cars will offer new viewing surfaces and more available time with which to target audiences. The ads will be location and identity based, enable in-vehicle transactions and purchases, and allow for interactive brand engagement. These cars will become mini entertainment centers - somewhat like your family room on wheels where you can be served up the best quality audiovisual entertainment.

AVs can also collect more data about the passengers and their environments. They have access to their locations, their routines and timing, current and past destinations, speed of travel, demographics and biometrics, present and future weather in their location, traffic conditions, and important dates and events relevant to them. With this knowledge, AVs can enable marketers to achieve unprecedented levels of precision in targeting ads to consumers.

You can be in your smart car and it knows from syncing with your calendar that tonight you are supposed to go to a neighbor's home for dinner. On your drive home from work, an in-vehicle ad suggests a nearby grocery store where you can pick up a dessert for the party, and you allow the vehicle to take you there based on the directions from the ad. In another situation you are in an AV and pass a chain coffee shop on the way to work every morning. Between the car's data and the store's data, they know that you like your coffee at this time of the day. You automatically get asked if you would like to stop to pick up a medium soy latte; if you say yes, the car is directed to the store, where your coffee is ready to be picked up at the counter, with the payment being automatically processed based on your identity.

You can see the confluence of benefits combining time of day with location with buyer behavior with real-time messaging. Now the ad, via the car, is not only responding to location but driving you to it - creating new opportunities for in-vehicle e-commerce.

Taking it one step further, the advertising industry could produce interactive, branded in-vehicle advertising or entertainment experiences. In the AV, companies could serve immersive augmented reality advertisements offering passengers the opportunity to feel as though they are interacting with products and thus creating a very engaging experience. These immersive advertisements put the consumer at the center of the ad - something that may incline them to connect more strongly with the brand.

Who Is Leading the Push into Media and Advertising in AV?

As at other inflection points in the industry, often the innovation comes from nontraditional media companies. Traditional media are looking at how to adapt to the availability of AVs as a new media channel. Newer companies - some yet to be formed - are likely to develop an entirely new approach to this new media device that will be highly disruptive.

In between those two extremes are companies like Google who are investing heavily in the development of AVs. Like their investment in smartphones, they are interested in providing and owning the platform that these devices will be built and run on. Just as they saw the potential for the smartphone as the next platform for advertising, they can foresee that the AV will be an even larger platform for advertising. They have been investing in Waymo for over 10 years with that goal in mind.

We also expect the traditional car companies to see advertising in AVs as an aspect of how they derive revenue. Since they are the manufacturing and service channel for these AVs, they can license application-programming interfaces, sell data to advertisers to lower the overall ownership costs of the vehicle, and derive revenue.

Impact on Advertising of Cars

For car brands, the themes of car advertising will shift significantly. Instead of promoting a car for its speed and features, AV companies like Tesla may promote cars based on the ease-of-use and life experiences they can provide.

It is likely that 15 years from now auto ownership will shift to MaaS and ownership and driving of cars will be limited to entertainment and leisure-time activities. In that scenario the car brands - potentially in partnership with ridesharing companies like Uber or Didi - behave like services companies and will market the experience rather than the product.

The Car as a Social Media Platform

If social media is a space where websites and applications enable users to converse and network, with increasing vehicle-to-vehicle communication, AVs could become the ultimate social media platform. Using a suite of sensors, AVs will be constantly talking to each other and their drivers about speed, direction, location, braking status, etc. Like a social media platform, they may be able to carry out all kinds of messaging - on behalf of the driver, potentially without the driver or passengers active participation.

What Is Happening Now?

The media and advertising industries are eagerly waiting for AVs to become mainstream so they can take advantage of the new opportunities for people to view media content and advertising. It will likely take many years to reach a level of autonomy such that people are able to completely take their eyes off the road and consume content.

Social, Legal, Ethical, and Business Impact

Like with all disruptive technologies, the current norms, practices, and laws will prove insufficient to handle this new media device:

- What will the car company control in terms of viewing experience relative to what you as the driver can access online?
- What are the privacy implications of all the detailed personal data that could be available for media companies to target advertising to the driver and passengers of the AV?
- At this time, drivers own their own data. If we move to MaaS, who will own the data of the person in the vehicle? Will it be the consumer, the vehicle manufacturer, or the ridesharing platform, or a combination?
- Who handles identity management so that only appropriate content is shown to the consumer in the vehicle?

One thing is clear - in the battle for media content advertising dollars, whoever owns the contextual data will ultimately decide who controls the content and ads, and who gets the revenue.

Recommendations

The marketing and advertising industry is in tune with the current automotive media use-models. With AV technology becoming pervasive, a precious asset - attention time of the driver - will become available. When combined with greater insight into the preferences of the occupants of the AV and their context, it will enable the next generation of media engagement.

1. Current Media: As AVs may obsolete the current use-models in marketing and advertising, exploring other uses for the existing assets in an AV-centered

world would be wise. Examples may include reuse of billboard assets for communication infrastructure.

2. Future Models: With more pervasive deployment of AV technology, there is opportunity for existing and new media companies to expand their targeted marketing/advertising offerings and create new business opportunities. These future engagement models are dependent on the solution to safety issues within the AV and will take some time to be fully deployed.
3. Mobility Integrated Business Models: Finally, unlocking the next generation of marketing/ advertising is quite valuable, so it would be wise to explore integrated business models which defray the cost of mobility through the use of advertising revenue. This could be especially beneficial in the context of public transportation.

Summary/Conclusions

The automotive ecosystem is very large. The utility provided by advanced mobility services has the potential to substantially disrupt the current situation. With the essays, each of the major entities has examined the current status quo and the challenges/opportunities. However, the extent of penetration for all the entities is fundamentally gated by the verification and validation (V&V) of AVs.

In Jim Boxold's essay, we learned about the challenges faced by state policy makers in areas such as basic statutory definitions of a driver, enabling regulatory structure to test AVs while maintaining safety, thinking through the updates required in process for items such as accident investigation, and finally, managing the potential investments required for the infrastructure to support CAVs. Central to these decisions is understanding clearly the operating space for AVs, the technology to measure safety with robustness, and viable business models. In Dale Neef's essay, he outlines the next level potential issues for local planning, finance, and law enforcement.

In Gary Ralston's essay, we learned about the large footprint the automotive ecosystem has on commercial real estate. With examinations of ideas around truck platooning, MaaS, and last-mile public transportation, Gary examined the impact on the use-models in commercial real estate. Central to the viability of these models is the solution to the vexing V&V challenges for AVs. Even if these challenges are solved, the shift in usage may move from less upfront consumption with dealerships but a dramatic increase in aftermarket/maintenance. Pankaj similarly outlines the current automotive ecosystem on media and advertising. AVs offer a new opportunity to access very precious attention cycles from consumers and sufficient data such that the interaction can be very targeted. However, all of this is also dependent on V&V robustness to the point of enabling this increased attention on nondriving tasks.

With the essay on legal frameworks, J.W. Taylor uses his vast experience in the legal aspects of transportation, he comprehensively examines the current legal structures which exist today and how liability may shift in the era of AVs. Obviously, since the fundamental test in today's legal framework is a "reasonable person," the critical question will be the expectation of AVs as defined by a clear view of their domain of operation. With a clear view of the definition of domains of operation and associated V&V testing for robustness, a legal framework for liability can be built. Richard Watt's essay dovetails directly into the legal framework as large auto insurers start examining the gradual shift from human to automated driving system (ADS) today and AVs in the future. They are gathering data to build actuarial tables and watching the legal environment to access liability.

Critical work to be done is to resolve the fundamental research issues in AV V&V, and until the issues are solved, clearly define the domains of operation which can set expectations for the various use-models enabled by AV technology. In this context, efforts such as SAE Automated Vehicle Safety Consortium are looking at standards around safety [14].

SAE EDGE™ Research Reports

SAE EDGE™ Research Reports, like the present report on "Unsettled Topics Concerning Automated Driving Systems and the Transportation Ecosystem," are intended to push further out into still unsettled areas of technology of interest to the mobility industry. SAE launches these reports before attempting to form a joint working group, let alone a CRP or a Standards Committee.

SAE EDGE™ reports are intended to be quick, concise overviews of major unsettled areas where vital new technologies are emerging. By definition, an unsettled area is characterized more by confusion and controversy than established order. Early practitioners must confront an absence of agreement. Their challenge is often not to seize the high ground but to find common ground. These scouting reports from the frontiers of investigation are intended merely to begin the process of sorting through critical issues, contributing to a better understanding of key problems and providing helpful suggestions about possible next steps and avenues of investigation.

SAE EDGE™ Research Reports, therefore, are fundamentally distinct from the more formal working groups approach and far removed from the more mature CRP and standard's development process. They begin by convening an informal meeting of subject matter experts working in an unsettled area of technology from industry, government, and the research community. A moderated but collegial discussion ensues regarding their shared interest. This conversation attempts to identify critical issues hindering a clear understanding of important topics, preventing a common agreement on potential solutions, or slowing progress in key areas of research.

An SAE EDGE™ Research Report is then drafted and circulated for comment to meeting attendees and other interested parties. After review, the final result of this process is a report like the current examination of unsettled issues concerning the sensors needed for automated vehicles.

Next Steps for Automated Driving Systems and the Transportation Ecosystem

This publication should be considered only as a first step toward clarifying the issues around ADS and the transportation ecosystem. The intention behind this and other SAE EDGE™ Research Reports is to start a dialogue among interested parties on important industry-wide topics that require further attention. The expectation is that these explorations of unsettled areas of technology will lead to the formation of working groups and, ultimately, committees that can address and resolve the issues they raise, producing a framework for developing a common vocabulary of definitions, best practices, protocols, and standards needed to support continued progress toward safer and more innovative automotive products.

The experts' meeting that gave rise to this publication demonstrated a great willingness on the part of the industry to define the terminology, procedures, and eventually the standards needed to enable ADS and the transportation ecosystem technology to move ahead as quickly and efficiently as possible.

This SAE EDGE™ Research Report on ADS and the transportation ecosystem identifies the following key topics for further pursuit, both through continued informal discussions among industry practitioners and through more formal working groups:

- AV impact on public policy
- How AVs will integrate into current commercial real estate and public infrastructure as well as future smart cities
- The determination of liability and the management of insurance claims regarding AVs
- The new media environment that will integrate with AVs

Recommendations

The overall recommendations of this SAE EDGE™ report can be summarized as follows:

1. Education: The automotive ecosystem is very large and AV technology is a large incremental shift for markets ranging from insurance to commercial real estate. It is vitally important that all the relevant markets build expertise in the basis of AV technology to understand, absorb, and manage the transitions.
2. Test and Validation Science: Whether it is public policy, insurance, or legal frameworks, the underlying gate are the technical elements surrounding the test and validation science surrounding AVs. This science will be the key kernel around which the other layers of capability will become viable. It is worthwhile for all the players in the ecosystem to cooperate to enable the development of this critical technology. Industry bodies such as SAE, working with other connected bodies, have an important role in this area.
3. Financial Modeling: Given education and some projections on the progress of test and validation science, all the major entities need to build modeling capabilities which stress test current assumptions based on the reality of AVs.

Abbreviations/Definitions

5G - Fifth-generation cellular network technology

ADAS - Advanced Driver Assistance System, also ADAS+. Advanced Driver Assistance System(s). Vehicle-based intelligent safety systems that can control steering, braking, and acceleration under certain circumstances. The human driver must continue to pay full attention (monitoring the driving environment) at all times to perform the rest of the driving task. ADAS are found in vehicles with Automated Driving Levels 1 and 2 as defined by SAE J3016™. Vehicles at Levels 3 and higher have ADS capabilities.

ADS - Automated driving system. Intelligent computer control of vehicles based on sensor input, characteristic of vehicles at L3 and higher *Automated Driving Levels*. See the entry for this term below.

Automated Driving Levels (as defined by SAE J3016™)

L0 - No Driving Automation. A human driver performs the entire Dynamic Driving Task (DDT). Other vehicle systems may provide warnings or support, for example, braking systems with automatic skid control.

L1 - Driver Assistance. Driving automation system is capable of executing either the longitudinal or the lateral vehicle motion control subtask (not both). It disengages immediately upon driver request. Human driver is present at all times to perform all control tasks not done by the automation system, to determine when it is appropriate to engage the automation system, to supervise all operations of the automation system, and to intervene as necessary to maintain safe vehicle operation. Example: automatic speed control ("cruise control") for highway driving.

L2 - Partial Driving Automation. Driving automation system is capable of executing both the lateral and the longitudinal vehicle motion control subtasks. It disengages immediately upon driver request. A human driver must be present and attentive at all times to perform control tasks not done by the automation system, to determine when it is appropriate to engage the automation system, to supervise all operations of the automation system, and to intervene as necessary to maintain safe vehicle operation. Total driver disengagement from human control mechanisms (hands off wheel and feet off pedals) is possible in narrowly limited driving scenarios, e.g., lane centering.

L3 - Conditional Driving Automation. A human driver must be present and attentive at all times to operate the vehicle outside the Operational Design Domain (ODD) of the Automation Driving System (ADS) and to take over control of the vehicle from the ADS within its ODD as necessary. The ADS can only be engaged in its ODD but can perform the entire DDT when engaged. It automatically disengages when it determines either that its ODD limits are about to be exceeded or it is experiencing a performance-relevant system failure, first, by issuing a timely request to intervene to the human driver and, second, by disengaging an appropriate time after issuing the request. It also disengages immediately upon driver request. The human driver is responsible for verifying the operational readiness of the vehicle, handling all DDT outside the ODD of the ADS, determining when the ADS can be engaged, and serving as the fallback-ready driver when the ADS is engaged, responsible for maintaining a minimal risk condition by actively intervening in ADS operation or assuming vehicle control on automatic ADS disengagement. An ADS ODD might include, for example, freeway driving or light traffic on open roads outside urban areas.

L4 - High Driving Automation. Human driver is responsible for verifying the operational readiness of the vehicle and determining whether to engage the ADS. The ADS will permit engagement only within its ODD. When the ADS is engaged, the human driver has no fallback role and may ride as a passenger. Once engaged, the ADS will transition to a minimal risk condition before it disengages (1) as the result of a performance-relevant system failure, (2) after an automatically issued request to intervene is ignored by a human driver, or (3) in response to a user request that it achieve a minimal risk condition. Optionally, a human driver may reassume DDT, by requesting the ADS disengage and waiting for it to turn over control of the vehicle once the ADS determines that conditions are appropriate; or, in a fallback role, following an ADS request to intervene; or, after the ADS has automatically disengaged, following achievement of a minimal risk condition.

L5 - Full Driving Automation. The ADS ODD expands to permit engagement under all driver-manageable on-road conditions. L5 vehicle operation, when the ADS is engaged, is identical to L4 vehicle operation.

AI - Artificial Intelligence. Computer systems that can exercise judgment and make decisions, for example, recognize objects from visual inputs, understand spoken or written words and translate between languages, make optimal choices between alternative moves relative to achieving a goal (e.g., playing chess, navigating)

AV - Autonomous vehicle

CAV - Connected and autonomous vehicle

EDR - Event Data Recorder

IoT - Internet of Things. IoT refers to the networking of everyday objects via inexpensive embedded chips with radio frequency identification capabilities (RFID).

MaaS - Mobility-as-a-service

NHTSA - National Highway Traffic Safety Administration. The federal agency charged with writing and enforcing federal motor vehicle safety standards.

MPO - Metropolitan Planning Organization

OEM - Original equipment manufacturer

RFID - Radio frequency identification

V2I - Vehicle-to-Infrastructure communication

V2X - Vehicle-to-Everything communication. Among other possibilities, "X" can be another vehicle, a pedestrian, roadside equipment, or the Internet.

V&V - Verification and validation

VMT - Vehicle miles traveled

Acknowledgments

Recognition should go first to all the participants, many of who also provided feedback on the draft version of this publication. Without their input and initiative, this SAE EDGE™ Research Report would not have been possible.

Jim Boxold, Florida Secretary of Transportation

Gary Ralston, Managing Director of SVN

J.W. Taylor, Transportation and Logistics Legal Specialist

Dale Neef, Expert in Local Government Planning for "Smart Cities"

Richard Watts, Sr. VP Progressive Insurance Ret

Pankaj Mayor, Marketing Expert, ex-CMO of public software companies

Additional thanks go to the following individuals who also had the opportunity to provide feedback on the draft version of this publication. Their contributions helped shape this publication.

Contributors

Dr. Mustafa Ilhan Akbas, Embry-Riddle Aeronautical University

Dr. Arman Sargolzaei, Florida Polytechnic University, Advanced Mobility Institute

Kristen M.J. Johnson, Esq., P.L.; Elizabeth "Elle" Slattery, Esq., and Wendy L. Fisher, Esq. - Taylor & Associates, Attorneys at Law, P.L.

The authors of this document together with the SAE Team responsible for its creation join in expressing our deepest appreciation to all the individuals mentioned above.

Dr. Rahul Razdan

Advanced Mobility Institute, Florida Polytechnic University

(external consultant to SAE International)

References

1. Razdan, R., "Unsettled Technology Areas in Autonomous Vehicle Test and Validation," June 12, 2019, https://www.sae.org/publications/technical-papers/content/epr2019001/.
2. Corwin, S. and Pankratz, D.M., "Forces of Change: The Future of Mobility," *Deloitte Insights*. https://www2.deloitte.com/us/en/insights/focus/future-of-mobility/overview.html, accessed October 22, 2019.
3. Sahawneh, S. et al., *Requirements for the Next-Generation Autonomous Vehicle Ecosystem* (Huntsville, AL: IEEE Southeastcon, 2019).
4. Soniak, M., "When and Where Was the First Car Accident?" *Mental Floss*, Dec. 6, 2012. http://mentalfloss.com/article/31807/when-and-where-was-first-car-accident, accessed October 22, 2019.
5. Kunkle, F., "Fatal Crash with Self-Driving Car Was a First - Like Bridget Driscoll's Was 121 Years Ago with One of the First Cars," *The Washington Post*, WP Company, Mar. 22, 2018, https://www.washingtonpost.com/news/tripping/wp/2018/03/22/fatal-crash-with-self-driving-car-was-a-first-like-bridget-driscolls-was-121-years-ago-with-one-of-the-first-cars/, accessed October 22, 2019.
6. Matthiesen et al., "DOC," American Bar Association - "Negligence," n.d., retrieved from https://www.americanbar.org/groups/public_education/resources/law_issues_for_consumers/everydaylaw0/health_care/personal_injury/negligence/, accessed October 22, 2019.
7. Wickert, G., "Contributory Negligence/Comparative Fault Laws in All 50 States," Matthiesen, Wickert and Lehrer, S.C., 2019, https://www.mwl-law.com/wp-content/uploads/2018/02/COMPARATIVE-FAULT-SYSTEMS-CHART.pdf (accessed Aug. 19, 2019).
8. Faegre Baker Daniels, "5 Defenses for Autonomous Vehicles Litigation," Insights, n.d., retrieved from https://www.faegrebd.com/en/insights/publications/2018/9/5-defenses-for-autonomous-vehicles-litigation, accessed October 22, 2019.
9. Reilly, P.H. et al., "5 Defenses for Autonomous Vehicles Litigation," Publications/Insights/Faegre Baker Daniels, https://www.faegrebd.com/en/insights/publications/2018/9/5-defenses-for-autonomous-vehicles-litigation, accessed October 22, 2019.
10. Minami Tamaki LLP, "Family of Driver Killed from Tesla Car Crash Hires Minami Tamaki LLP to Explore Legal Options," Apr. 11, 2018, https://www.minamitamaki.com/2018/04/family-of-driver-killed-from-tesla-car-crash-hires-minami-tamaki-llp-to-explore-legal-options/, accessed October 22, 2019.
11. Korosec, K., "Volvo Will Accept All Liability When Its Cars Are in Autonomous Mode," *Fortune*, Oct. 8, 2015. https://fortune.com/2015/10/07/volvo-liability-self-driving-cars/, accessed October 22, 2019.
12. Ryskamp, D.A., "Self-Driving Cars Will Likely Increase Product Liability Litigation," *The National Law Review*, n.d., retrieved from https://www.natlawreview.com/article/self-driving-cars-will-likely-increase-product-liability-litigation, accessed October 22, 2019.
13. BBC, "UK Wants Fully Autonomous Cars on Road," *BBC News*, Feb. 6, 2019. https://www.bbc.com/news/technology-47144449, accessed October 22, 2019.
14. SAE International®, "SAE International, Ford, General Motors and Toyota Announce Formation of Consortium to Address Autonomous Vehicle Safety," Apr. 2, 2019, https://www.sae.org/news/press-room/2019/03/sae-international-ford-general-motors-and-toyota-announce-formation-of-consortium-to-address-autonomous-vehicle-safety.

Contact Information

EDGEresearch@sae.org.

The authors of this document together with the SAE Team responsible for its creation join in expressing our deepest appreciation to all of the individuals who contributed.

www.ingramcontent.com/pod-product-compliance
Lightning Source LLC
LaVergne TN
LVHW070941210926
848482LV00003B/275